Contents

Foreword

The story of the rotary aero engine is a very unusual one. When it was invented around 1908 it took the world of aviation by storm, despite being very expensive. It was clearly so much better than any other type of engine that it sold like the proverbial hot cakes. Tens of thousands were made in the First World War, output climbing each year until the Armistice. Then, exactly ten years after its introduction, the rotary disappeared as an industrial product almost overnight.

This is not because its qualities were in any way specifically geared to military aviation, and neither because of the world surfeit of unwanted wartime production. The rival water-cooled inlines and air-cooled static radials continued to be built, for air forces and the infant airlines; it was only the rotaries that disappeared from the factories.

Many reasons have been put forward to explain how it was that something that burst on the scene like a nova should have had such a short life. They revolve around lubrication problems (and especially the use of castor oil for this purpose), windage losses in spinning the whole engine round, the difficulty of gearing down the drive to the propeller, centrifugal stresses, gyroscopic influences on the aircraft and many more. In this well-reasoned book Andrew Nahum shows that the real reason, at least the crucial one, was more fundamental.

The subsequent story of the (static) piston aero engine continues to be a fascinating one. Even today the story is far from over. Air cooling or water? Horizontally opposed or radial? Reciprocating or Wankel-type 'rotating combustion'? Otto cycle or diesel? The one thing we can be sure of is that the engine itself no longer rotates!

Bill Gunston

SCIENCE MUSEUM

The *Rotary* AERO ENGINE

Andrew Nahum

LONDON   HER MAJESTY'S STATIONERY OFFICE

© Copyright The Trustees of the Science Museum 1987
First published 1987
ISBN 0 11 290452 1

Units

The data in this book has been drawn largely from contemporary British sources. These use Imperial units for British, French, and captured German engines. It has been decided not to convert these figures into metric units. Specific power outputs, however, are given as brake horsepower/litre. Although this hybrid is being replaced by kilowatts/litre, the older term has been used for a long time in Europe, and has the advantage of readily conveying an impression of the power output and efficiency of an engine to the broad group of people with an interest in internal combustion engines.

front cover Louis Paulhan with his Voisin and 50 hp Gnome. Reims, 1909.

back cover The Caudron Aircraft Company operated flying schools in both Britain and France before the First World War. The 50 hp Gnome was a favoured power plant.

Preface

The Gnome quickly became the favourite engine for pioneers. Flying personalities at Hendon in 1913, from contemporary postcards.

The aircraft rotary engine came to prominence in 1908 with the development of the 50 hp Gnome. This engine set new standards for power and light weight among aircraft power plants. It was adopted by many pioneer aviators and used extensively to set records for endurance, speed and height. It was soon followed by more powerful Gnome models, and by variants from other manufacturers, principally Clerget and Le Rhone. The type was used extensively by the air forces of both sides during the First World War and was developed to give up to 200 horsepower. However, by 1918, rotaries were being eclipsed by new types of stationary engine and virtually all manufacture and design ceased after the war.

Various explanations have been advanced to explain the rapid demise of the rotary. Those most frequently cited are the flight control problems caused by the gyroscopic effect, high fuel consumption and the additional structural loads imposed by rotation of the crankcase and cylinders. In this book, it is argued that none of these explanations is adequate and that the real limit to rotary development was a 'torque ceiling' which was a fundamental consequence of the way in which the engine works. The reasons for this are discussed and a comparison is made of power and torque test figures between rotaries and stationary contemporary types to support the argument.

Acknowledgements

The author would particularly like to thank John Bagley, Curator of the National Aeronautical Collection at the Science Museum, for his encouragement and for many helpful suggestions. Alfred Bodemer provided useful insights into the early French industry, and was generous with photographs from his collection. Photographs were also provided by The Imperial War Museum, the Royal Air Force Museum, the Musée de l'Air, the Conservatoire National des Arts et Métiers, the Ministry of Defence Air Historical Branch, the Smithsonian Institution and Captain D. W. Phillips. Other photographs are drawn from Science Museum archives. Thanks are due to Jonathan Cape Ltd for permission to reproduce the extract from *Winged Victory*.

Henri Farman (right) and 50 hp Gnome. The aircraft is Claude Grahame-White's Farman, rigged at the airship hangar, Wormwood Scrubs, awaiting his attempt to win the Daily Mail £10,000 prize for a flight from London to Manchester.

Introduction

The rotary aero engine sprang into prominence with the introduction of the 50 hp Gnome in 1908, and in particular, with Henri Farman's adoption of this model for the famous 1909 Rheims meeting, 'the first great aviation meeting of history', where he carried off the Grand Prix for the greatest non-stop distance flown (180 km) and set up a world record for endurance. The rotary quickly became the aeronautical powerplant *par excellence* and dominated records and results in the remaining years of peace.

During the First World War, the rotary was initially the most important type of aircraft power unit on the Allied side, although towards the end of the struggle the relative contribution began to decline as competitive stationary engines entered service. However, developed examples of the rotary stayed in front-line use up to the end of hostilities. By 1919, virtually all production and design work on rotaries had stopped, and thereafter, few references are to be found to them in aeronautical and engineering journals.

The rotary is a special form of air-cooled radial engine, in which a number of cylinders are arranged like the spokes of a wheel, around the crankshaft. For an aero engine, the radial arrangement has several advantages. Grouping the cylinders in one plane around a single crank gives a very short engine, with a considerable weight saving over the in-line type, in

A 50 hp Gnome on test at the factory on a portable 'gun carriage' test bed.

which every cylinder requires its own crank 'throw'. Furthermore, the cylindrical crankcase of the radial is an efficient 'drum' type of structure which resists engine working stresses by radial tension loads, and thus can be built light. In the rotary version, for better cooling, the mode of operation is reversed from that of the normal radial, so that the crankshaft remains stationary (and supports the assembly), while the cylinders, and the body of the engine rotate around it. Perhaps this counter-intuitive (some would say perverse) arrangement explains the continued fascination the species holds over aviation historians and enthusiasts. (A French author writing in 1910 commented; 'their bizarre appearance, the originality of their parts, draws and retains the attention'.) In addition, the engines themselves are superbly constructed, typically of fully machined fine-section nickel-steel components, made and fitted to the highest standards of Edwardian engineering. Given this abiding interest, it seems remarkable that there appears to have been no convincing attempt to explain the most curious and absorbing features of the 'rotary phenomenon' – the dramatic and immediate superiority over all competing types at its inception, and the equally abrupt disappearance from the world of aviation in 1918.

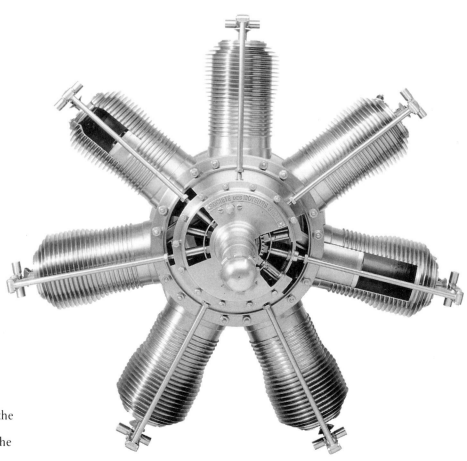

Sectioned 50 hp Gnome in the Science Museum (serial number 829) and detail of the cylinder head.

PART ONE

Triumph and Decline

The Hum of the Fifty Gnome

Give me the hum of the Fifty Gnome,
    Flying o'er hill and dale,
Taking me whither I care to roam,
    To uplands, or sun-kissed vale,
Take me away from the busy street,
    Away from the township's toil:
Give me the hum of the Fifty Gnome –
    The taste of the castor oil.

Give me the hum of the Fifty Gnome,
    Give me the chance to go
Over the city's cross-topped dome,
    Over the hills of snow.
Never was loved one half so dear,
    Never was voice so sweet,
Give me the hum of the Fifty Gnome –
    The feeling of icy feet.

by 'The Dreamer'
Flight, 28 June, 1913

The Introduction of the Rotary into Aviation

For his first flights in 1908, A.V. Roe's triplane (below) used a modified 6 hp V-twin JAP, originally designed for motorcycle use.

For the pioneer aviator, the choice of power unit was between automobile derived water-cooled engines, which were heavy, and air-cooled cyclecar or motorcycle units which were prone to suffer from overheating. For example A. V. Roe first flew using an adapted JAP V-twin cylinder motorcycle unit of limited power. (One exception was the impressive purpose–designed V8 Antoinette engine.)[1] However, the development in 1908 of the Gnome, the first successful aircraft rotary engine, made it the premier choice for aviators. The engine was the work of the three brothers Seguin – gifted engineers, whose grandfather, Marc Seguin had constructed the first French locomotive in 1829, and had pioneered both the horizontal fire-tube boiler for locomotives and suspension bridge construction. (It is noteworthy that the family were involved continuously in engineering over several generations. They were also descended indirectly from the first aeronauts, Joseph and Étienne Montgolfier.)

In 1906, Louis Seguin, the eldest brother, had formed the Société des Moteurs Gnôme to build stationary industrial engines. (The name Gnome was chosen to suggest 'a diminutive, sturdy little worker'.) Louis was joined by his brother Laurent, who designed a rotary engine 'specifically for the requirements of contemporary aviation'. His interest was doubtless stimulated by the considerable amount of discussion at this time in Paris on aviation, and about *'le problème du moteur léger'*. (In 1905, the aviation

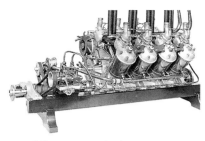

'The engine which got Europe off the ground'. The water-cooled Antoinette was the first commercial purpose-built aero engine.

propagandist Ernest Archdeacon had written to the Wright brothers requesting them to give a demonstration, adding 'if you are our masters in aviation, we are certainly yours in the matter of light motors'.) Augustin Seguin, the youngest brother, recalled in 1965, 'My oldest brother Louis Seguin, director of the Société des Moteurs Gnôme took his younger brother Laurent in with him as an engineer. Laurent had the idea of constructing a rotative motor for the developing field of aviation, and designed it himself, very light, and weighing a kilogram per horsepower. My two brothers collaborated on the construction of this motor, which was to become famous, and decided with common accord not to separate their names in the invention, but it would be unjust not to give each his due.'

The first experimental engine was a 5-cylinder model which was said to develop 34 bhp. It has always been assumed that this was a rotary, but recent research by Alfred Bodemer has drawn attention to a 5-cylinder static radial at the Musée de l'Air, believed to date from 1906, which appears to be the prototype. This motor is said to have been used in a high

This five cylinder *static* radial appears to be the prototype Gnome aero engine. French sources indicate it was built in 1906.

speed motor boat, the 'Isabelle Gnome', though this does not argue necessarily against it having been designed for aeronautical use. Like aviation, motor boat racing appealed to a small, rich clientèle and the two worlds were inter-connected; some engine makers (especially Antoinette) sold engines for both the new sports.

It seems reasonable to suppose that the Seguins then forsook the static radial in favour of the rotary in the search for better cooling, for an early experimental 7-cylinder Gnome exists at the Conservatoire National des Arts et Métiers in Paris.[2] Curiously, this has no cooling fins. It is unclear whether the Seguins believed that rotary motion itself would be sufficient to cool the engine, or whether it was simply built for short runs, to prove

This experimental unfinned Gnome may have been constructed for brief runs only, to test the validity of the rotary principle.

Rigging Grahame-White's Farman for the London-Manchester attempt. Fitters from the Gnome factory followed in a tender car to look after the motor.

the feasibility of rotary operation, the fins being omitted to speed production and reduce machining costs. The first production type, the 7-cylinder 50 hp 'Omega', was first shown at the Paris automobile show of 1908. In the spring of 1909, the inventor M. Ravaud fitted one to his Aéroscaphe, an unsuccessful hydroplane/aircraft, entered in both the motor boat and aviation meetings at Monaco. However it was probably Henri Farman's use of the Gnome at Rheims later in the year which launched its commercial success. There were also notable flight achievements by Louis Paulhan, who seems to have been provided with his engine by the Seguins for promotional purposes.

The rapidly growing demand for the engine was satisfied efficiently – presumably because of the pre-existing industrial structure built up by Louis Seguin for the stationary engine business. The production figures for the years before the outbreak of war illustrate the rapid expansion of the aero engine side. Some 4000 engines were constructed and the labour force increased to number a thousand. This was paralleled by an exceptional financial performance that makes some of today's high-tech stocks look tame! Out of net profits for 1911 of £149,000, the directors paid a dividend of 26 percent and repaid the entire share capital of £48,000. Profits for 1912 were up nearly 30 percent at £210,000, and after allowing for normal depreciation, £100,000 was taken to reserves and a dividend of 150 percent paid on the nominal value of the share capital. Much of this success must

Machining Gnome cylinders from solid.

Laurent (left) and Louis Seguin.

have been due to the energy and acumen of Louis Seguin. In 1912 he received the Cross of the Légion d'Honneur 'for his exceptional services in relation to the famous Gnome engine and its contribution to aviation'. The report in L'Aérophile commented; 'entirely devoted to his work, this eminent engineer and enlightened and active industrialist refuses to be a socialite, but lets his achievements speak for themselves . . .'

There were soon Gnome subsidiaries in most industrialized countries although the engines were expensive. In the first commercial aviation catalogue, that of the Aeroplane Supply Company for 1909, the 50 hp Gnome was priced at £520 (about the same as a good-quality medium-sized car, such as a Daimler or Vauxhall), while the 60 hp Belgian Vivinus was £280 and the English Green 60 hp, £365. However the power to weight ratio of the Gnome was superior to practically all other engines available at the time (3.6 lb/hp against 6.1lb/hp for the aircooled V8 Renault) and the price reflected its engineering refinement, and particularly, the high costs involved in production. (For example, the cylinders were machined from solid steel billets.) It is clear that a Gnome conferred greatly increased performance on early aircraft and was a major stimulus for the development of aviation. (After his epic channel crossing, Blériot is alleged to have proclaimed 'More power, more power. I'm going to get a Gnome'.) In 1913, Major J. D. B. Fulton, instructor at the Royal Flying Corps school wrote, 'Taking the good with the bad, it is impossible to refuse the Gnome a place in the very front rank of aerial engines. It may be said without fear of contradiction that this engine has made aviation what it is. . . . A careful study of its characteristics, therefore, will be well repaid, for no motor responds more readily to intelligent treatment and none can be more stubborn in ignorant hands.'

The Gnome factory offered a comprehensive parts service and could also supply aviators with the services of a fitter at fixed rates, plus hotel and second class rail travel.

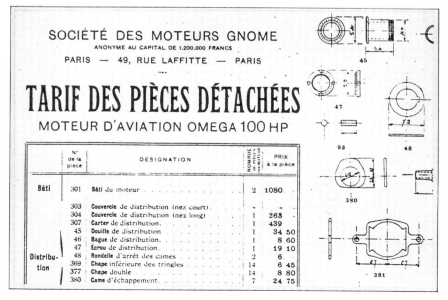

The 50 hp Gnome can be considered as the archetypal aeronautical rotary. The essential features that it embodies are; rotation of the cylinders and the crankcase (on which the propeller is mounted) around a stationary crankshaft and the aspiration of the petrol/air mixture through the hollow crankshaft itself. Thus the crankcase is utilized as a distribution chamber and from it, mixture passes to the combustion chambers through automatic valves in the piston crowns. These valves were always a weak point and adjustment was critical to make sure that the valve in each piston was loaded to come off its seat at the same pressure. 'Ace' aviators were said to have had their Gnomes stripped down overnight at air displays, particularly to check this adjustment. However, by the standards of early aero engines, the Gnome was not particularly temperamental, and was credited with being the first to be able to run for ten hours between overhauls. The selection of the Gnome by prominent flying personalities underlines its qualities, for they had a great deal at stake. The better pilots were able to earn extremely large sums from a combination of payments by aeroplane constructors, appearance money at aviation events, and prizes. In 1911,

Prier Monoplane, built by Bristol, 1911.

Gnome trade literature, 1911.

Louis Paulhan was said to have one of the biggest personal incomes in France – over a million francs, while the aviator Garros won 70,000 francs in a comparatively minor two day event, the *Circuit d'Anjou*. At the time, the yearly income, for instance, of a middle-ranking army officer might be 2000 francs. Most of the records for altitude, endurance and speed in the enthusiastic flying era that preceded the First World War were won by aircraft equipped with Gnome engines. There were clearly strong reasons for choosing a Gnome.

Many aviators carried a supply of postcards to autograph at events. Galy, with Caudron biplane, at Hendon in 1912.

Underside of piston *(cutaway)* showing coil springs.

In 1913, the Seguins introduced a new design, the Monosoupape, which eliminated the valve in the piston crown, and substituted transfer ports of the type familiar on two-stroke engines. (It is important to stress that though some experimental two-stroke rotaries were built, the Gnome 'Mono' and all the other significant makes operated on a normal Otto cycle.)

Important rotary engines were also devised by the Le Rhone and the Clerget companies (also both founded in Paris) in which conventional pushrod-operated inlet and exhaust valves were situated in the cylinder head. Like the Gnome, these employed the same principle of drawing mixture through the crankshaft and used the crankcase as a distribution

The 'automatic' inlet valve in the piston of the 50 hp Gnome is held on its seat by light coil springs. These often gave trouble. Factory mechanics could strip an engine down for inspection in two hours.

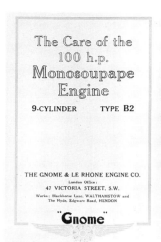

The Care of the
100 h.p.
Monosoupape
Engine

9-CYLINDER   TYPE B2

THE GNOME & LE RHONE ENGINE CO.
London Office :
47 VICTORIA STREET, S.W.
Works : Blackhorse Lane, WALTHAMSTOW and
The Hyde, Edgware Road, HENDON

"Gnome"

Title page, Gnome 'Mono'
instruction manual, 1917.

chamber, although in these engines mixture fed through individual manifolds from the crankcase up to each cylinder. (A more detailed description of the way a rotary engine operates is given elsewhere in the book.)

The success of the Gnome and its derivatives was widely attributed at the time to good cooling, and contemporary studies indicated that a comparable static radial engine would have to be given a forward speed of 30 mph to be cooled to the same degree as a rotary running on a test stand. It seems probable that the better distributed cooling which resulted from the combined windmilling of the engine and the propeller slipstream was an important element, and that it was this factor that allowed the engines to be built so light, for they operated at the limits of permissable thermal distortion, and employed extremely fine sections in most components.

The inlet manifolds, running from the crankcase to each cylinder head on this Siemens-Halske Sh III are a typical feature of later rotaries.

18

The Derivation of the Gnome Engine

The Seguins did not invent the rotary principle. However, they did adapt it to produce a purpose-built aviation engine that conferred greatly enhanced performance on the 'tentative' aircraft of the day. Among the various precursors, mention should be made of the rotary engines made by Stephen Balzer of New York for the experimental flying machines of Professor S.P. Langley, Secretary to the Smithsonian Institution. (With what has been described as 'a curious disregard for etymology', Langley named these craft 'aerodromes'.) Balzer's engines were converted from rotary to static radial operation by Charles Manly, Langley's talented engineering assistant. Another American rotary was the Adams-Farwell which was being manufactured for automobiles by 1901. Emil Berliner, the phonograph pioneer sponsored its development as a lightweight aero engine for his unsuccessful helicopter experiments. However Adams-Farwell engines did power fixed wing aircraft in the USA from 1910.

The rotary built by Stephen Balzer in 1900 for S.P. Langley's 'Aerodrome' experiments.

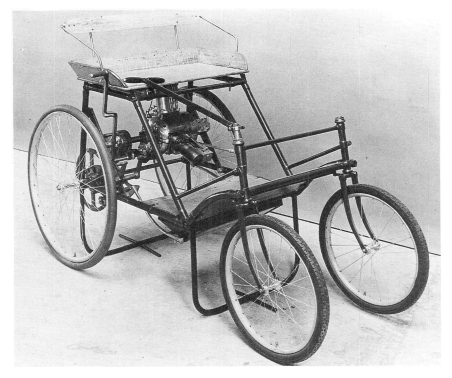

Stephen Balzer trained as a watchmaker at Tiffany's in New York, before he devised this three cylinder, rotary engined car in 1894.

19

Felix Millet patented his rotary, built into a cycle wheel, in 1888. It enjoyed modest success, and was manufactured by the Darracq company.

It has often been asserted that the Gnome design derived from the Adams-Farwell. Another Seguin brother, Marc, was French Consul in St Louis and could have reported to his family on the engine. An Adams-Farwell car is also said to have been demonstrated at the French Army manoeuvres in 1904. However it seems far more likely that the Seguins drew their inspiration from types they would have seen in Paris. At the Paris Universal Exposition of 1889, Felix Millet exhibited his 5-cylinder rotary engine built into a cycle wheel. (The best position for the engine in a motorized cycle was far from established at the time.) Millet had patented the engine as early as 1888, so must really be considered the pioneer of the practical internal combustion rotary engine. One example took part in the historic Paris-Bordeaux-Paris race of 1895 and the system was put into limited production by the Darracq company in 1900. It is also intriguing to note that at the same 1889 Paris Exposition, the German Daimler company exhibited an engine with auxiliary inlet valves in the piston crown. No doubt the Seguins learned of the Adams-Farwell and Balzer/Manly engines, but as engine builders they must have studied the exhibits on display at the various Paris exhibitions with great attention and the Millet would have been the rotary with which they were most familiar. The clue about the inlet valve in the piston is also intriguing, for this was to become a particular feature of the 50 hp Gnome. By contrast, the early Adams-Farwell engines had exhaust and inlet valves conventionally located in the cylinder head. Thus the main elements of the 'Gnome ensemble' were known in Paris at the time of its development. Furthermore, in 1899, the famous De Dion-Bouton company produced an experimental four cylinder rotary engine, specifically intended for aviation, though it was not fitted to an aircraft.

It should be borne in mind that Paris was perhaps *the* creative centre of internal combustion engineering for motive power at the turn of the century particularly in the field of lightweight, high speed power plants. It is noteworthy that when Langley and Manly became disillusioned with the progress being made by Balzer on the 'Aerodrome' motor they visited engine builders in Europe, and particularly Paris. However, at that time, these builders 'did not consider it possible to construct an engine of 12 horsepower weighing less than 100–150 kilograms, or that if they had thought it possible they would already have built it'. Manly paid particular attention to the engines on display in Paris and visited all the important manufacturers. Langley bought an air-cooled single cylinder De Dion-Bouton engine from the factory which was tested later at the Smithsonian by Manly. He adopted two features from it; the lightweight piston design, and the ignition.

Manly's final engine seems to have been outstandingly successful. It gave 52 hp continuously, from a weight of only 125 lb. Unfortunately the Aerodrome was less effective, and when tests were terminated in 1903, Manly's unique engine was set aside and led to no successors. It was left to the brothers Seguin to produce, in the Gnome, the first really light, powerful engine for general sale.

Arguably, the greatest contribution of the Seguins to the aeronautical science of their age was not the application of the rotary principle *per se* but a quite original look at the technique of constructing engines to minimum weight. In fact, the power developed was low on the basis of horsepower per litre, even by the standards of 1909 contemporaries (a mere 6.3 bhp/litre for the 50 hp Gnome), for the engine was optimised according to a different criterion – that of power to weight ratio. They used the highest strength material available – the recently developed nickel steel alloys – and controlled the weight by machining strong components from solid billets to very fine sections. The cylinder wall thickness of a 50 hp Gnome is a mere 1.5 mm and components such as connecting rods are good examples of intelligent design giving stiffness and strength in a light structure (longitudinal webs are created by milling deep central channels in the rod). It is

More powerful double row Gnome engines established several aviation records before the First World War. They were complex, and were supplanted by bigger single row types. This is a 200 hp 'Delta-Delta'.

The experimental Dunne D.7 bis monoplane (1911) with 70 hp Gnome.

interesting that no rotary manufacturer of note used aluminium for major structural components until 1917, in the Bentley BR 1, although the metal was used extensively to replace cast iron in more substantially built fixed engines. Contemporary aluminium castings were frequently flawed, and only during the war itself did high strength aluminium alloys that could be forged become available.[3]

From about 1911, the French army started to recognize the military potential of aircraft, and experimented successfully with their use for reconnaissance and gunnery direction during manoeuvres. The good performance of the Gnome and other rotaries led to their widespread adoption by the emerging air forces, while the main defects of high cost and a requirement for skilled attention could be tolerated in a military role.

By 1912, over 200 aircraft were in service in France, and Germany, careful to maintain parity, had slightly more. In contrast, Britain had a late start, and by this date possessed very few operational aircraft. The chairman of the Defence Sub-Committee on aviation drily remarked 'At the present time we have, as far as I know, of actual flying men in the Army about eleven, and of actual flying men in the Navy about eight, and France has about two hundred and sixty-three, so we are what you might call behind.'

Thereafter more attention was paid to developing British military aviation, but it has been said that at the outbreak of war, no British service aircraft was flying with a British-built engine, and only one aero engine of any type (the Sunbeam) was actually being built in this country. From the beginning of the war, in August, to the end of 1914, British companies produced only 193 aircraft, while 139 aero engines were procured – one third of these coming from France. The enormous ensuing increase in aircraft manufacture is one of the most remarkable industrial phenomena of the war, for in 1918, Britain produced 30,671 aircraft and procured 31,269 engines. Surprisingly, about one third of these still came from

The Dunne D.8 with a member of the Royal Flying Corps.

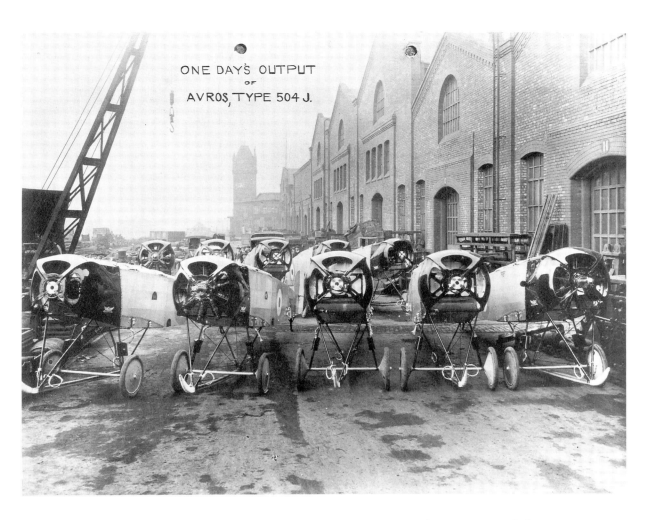

ONE DAY'S OUTPUT
OF
AVROS, TYPE 504 J.

The Avro 504J was built in large numbers, as a trainer, from 1916.

French makers. However, a major contribution to British domestic production throughout the whole war came from French rotary engines built under licence. A variety of engineering companies built these designs, though most had not previously been in the regular business of building engines. Throughout the war, the proportion of rotaries relative to other engine types (water- and air-cooled stationary engines) remained steady at about one-third of total production in each war year. The importance to the war effort of this French design contribution should be stressed, for the rotaries tended to be fitted mainly to aircraft types where a high power to weight ratio was needed, and so made a greater contribution to the development of the high performance 'fighting scout'[4] and to eventual Allied air superiority than is suggested by the numbers produced.

Major Rotary Engine Manufacturers

The success of the Gnome inspired many other manufacturers and inventors to experiment with the rotary, some with a degree of ingenuity that bordered on the eccentric. These designs were insignificant, both in terms of production and of their significance to aviation; they are not dealt with in this book. However, there were two other major manufacturers in Paris; Le Rhone (founded in 1912) and Clerget (1911). With Gnome, these concerns and their licensees supplied the bulk of Allied requirements during the war. (The principal licensee manufacturers in Britain were; for Gnome, Peter Hooker of Walthamstow; Le Rhone, W. H. Allen, Bedford; and Clerget, Gwynnes of Hammersmith.)

Le Rhone

The Rhone design was the work of the engineer L. A. Verdet, who was previously at Peugeot. Verdet also designed the famous Lion Peugeot racing

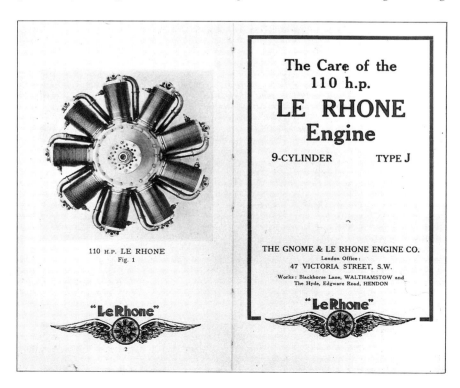

110 H.P. LE RHONE
Fig. 1

The Care of the
110 h.p.
LE RHONE
Engine

9-CYLINDER   TYPE J

THE GNOME & LE RHONE ENGINE CO.
London Office:
47 VICTORIA STREET, S.W.
Works: Blackhorse Lane, WALTHAMSTOW and
The Hyde, Edgware Road, HENDON

"Le Rhone"

Le Rhone instruction manual, 1918.

The Le Rhone caged big end assembly.

right Diagrammatic view from the Le Rhone instruction book.

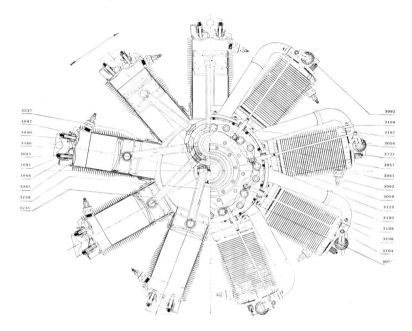

voiturettes. These highly successful cars were the last word in exploiting to the letter a racing formula that limited bore size (but not engine capacity). With an enormous stroke of 280 mm for a bore of only 80 mm, the top of the Lion V-twin engine was on a level with the driver's head!

The Rhone valve design was more conventional than the Gnome in that it employed normal inlet and exhaust valves in the cylinder head. These were operated by a single rocker and a push-pull actuating rod. This system was satisfactory, though it did preclude the use of any overlap for valve timing. Le Rhone steel cylinders had a pressed-in cast iron liner to produce a better running surface. The big end construction was unusual and used a system of slipper ended connecting rods which were captive in tracks in a cage or thrust block, which formed the outer part of the big end bearing. This removed the imbalance arising from the inequalities in piston acceleration that occurs between cylinders with a Gnome type master rod system and produced a smoother running engine.*

The Rhone company was taken over by Gnome in 1915, to form the Société des Moteurs Gnôme et Rhône. However the new concern produced the two distinct types throughout the war, and from their own factories supplied more than 20,000 engines.

---

*'My engine's been rotten since I over-revved it when we chased a Halberstater that got away'.
'Mine's a peach. It's a genuine Le Rhone Le Rhone'.
'The French are damn good at making engines . . .'
(from *Winged Victory* by Victor Yeates. This autobiographical novel gives a special insight into the flying qualities and operation of rotary-engined aircraft. A longer extract from this is given in the Appendix.)

SOCIÉTÉ DES MOTEURS

GNOME & RHONE

GNOME : 150 HP.
RHONE : 80 HP ; 120 HP ; 150 HP.

3 — Rue La Boétie — 3
PARIS

The Le Rhone company was taken over by Gnome in 1915.

Clerget

The Clerget was devised by Pierre Clerget, a manufacturer of industrial and automotive engines and with a Monsieur Blin, the Clerget-Blin company was established in Levallois-Peret, a north-western suburb of Paris well known for its numerous engineering workshops. The company also constructed some stationary aero engines. It is interesting that an example with the then advanced and unusual feature of aluminium pistons was displayed at the Paris aeronautical exhibition of 1909. This must be one of the earliest recorded uses of aluminium in this application. (W.O. Bentley recorded that he first saw an aluminium piston in 1913.)

Pierre Clerget (right) with a stationary Clerget. The pilot and constructor Marcel Hanriot is on the left.

right Nine cylinder 150 hp Clerget.

NOTICE

CONCERNANT

LE MONTAGE, LE RÉGLAGE
ET L'ENTRETIEN

DU MOTEUR ROTATIF

Type 9 B
130 HP

CLERGET-BLIN & Cie

37, Rue Cavé
LEVALLOIS-PERRET (Seine)
Terminus ; Tramway Madeleine-Levallois

Téléphone { WAGRAM 62-83.
{ WAGRAM 84-17.

Télégraphe : CLERGET-BLIN LEVALLOIS-PERRET

Title page, Clerget instruction book, c1918.

The Clerget was perhaps the most pragmatic of all the rotary designs (A Royal Flying Corps fitters' notebook preserved in the Science Museum has the comment 'best rotary' entered against the Clerget). The valve gear consists of inlet and exhaust valves in the cylinder head, like the Le Rhone, but these are separately actuated, allowing more flexibility in the choice of valve timing. Of the French rotary makers, only Clerget succeeded in bringing engines approaching 200 hp into service towards the end of the war.

Bentley

W.O. Bentley with one of his 1922 Tourist Trophy team cars, Brooklands 1922.

With the outbreak of war, W. O. Bentley was installed by the Admiralty into the design department of Gwynnes. (The Gwynnes-built Clergets were being used by the Royal Naval Air Service).

He was instrumental in making various design improvements (including the well known adoption of aluminium pistons) but 'was soon neck-deep in the politics, manoeuvrings and jealousies that arise when an outsider is let into the design department'. The Admiralty then moved Bentley to the Humber works at Coventry and commissioned him to design a new rotary. This was a remarkably courageous and prescient decision. Though Bentley was successful as an engine tuner and competitor in pre-war motor sport, he had no *bona fides* as a designer of complete engines. The 150 hp BR 1 (originally designated the AR 1 or Admiralty Rotary) was used to good effect in versions of the Sopwith Camel. The cylinder construction was one that Bentley had pressed for unsuccessfully at Gwynnes, and used aluminium cylinders for good heat conductivity, with cast iron liners shrunk in.

Apogee of the rotary: the 200 hp Bentley BR 2.

This effectively cured the problem of asymmetrical thermal distortion in rotary cylinders. Each cylinder was retained by four long studs running from the cylinder head to the crankcase – an excellent design feature. The BR 2 was an enlarged development of the BR 1, which went into production in spring 1918. Rated at 200 hp, it had a maximum output of over 230 bhp (though it was only 93 lb heavier than the BR 1) and was intended by the designer to leapfrog any foreseeable enemy developments. The test results made a powerful impression on the Air Board, and a production target of 1500 engines per month was suggested. This programme was too large for Humber alone, and the Daimler company was made principal contractor, with Humber and Crossley also contributing. The BR 2 was the most powerful rotary to see service and could be said to represent the swan-song of the aircraft rotary engine.

Bentley rotaries being manufactured and tested during 1918 at the Humber and Crossley works.

Siemens and Halske

Differential gearing (sectioned) of the Siemens-Halske Sh III.

Rotary engines were produced during the First World War by the Siemens and Halske division of the Siemens company. The other division, Siemens-Schuckert, built airframes.

The distinctive feature of these engines was the contra-rotation of the engine body and the crankshaft, which is produced by a differential gear mechanism. The 1914 patent was taken out in the name of the Siemens company and C. B. Burdon, who may have been the designer. Another source has named the designer as Franz Dinslage.

The culmination of the company's efforts was the Sh.III, a 200 hp engine with a swept volume of 18.6 litres. This was ordered in considerable quantities and in addition to the company's own production at Siemenstadt, Berlin, the firm of Rhemag at Mannheim was constituted in 1917 to build the engine under licence, with an initial order of 1000 units. By the end of the war some 1200 engines had been produced. The Sh.III was fitted to the Siemens-Schuckert D III and D IV scouts. These aircraft had an exceptional rate of climb, and have been described as 'the best German fighters to reach operational status'.

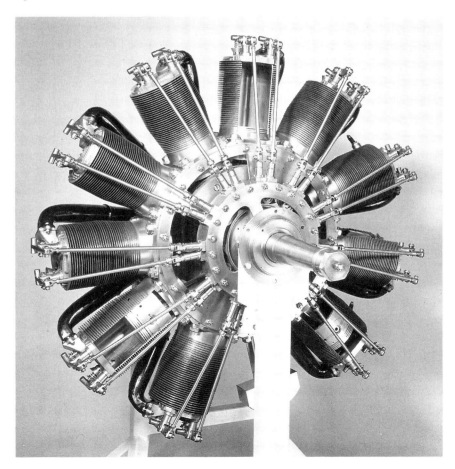

The contra-rotating Siemens-Halske Sh III. An alternative path to that used by W.O. Bentley in the search for high power.

Oberursel

Before 1914, the Oberursel company of Frankfurt had obtained a licence to build Gnome engines. However, these pre-war designs became obsolete, and with the encouragement of the Army's Inspectorate of Aviation, the company embarked on the manufacture of a copy of the 110 hp Le Rhone. This is striking, for Germany had pioneered the high efficiency water-cooled in-line engines which could be regarded as the progenitors of mainstream liquid-cooled aero engines subsequently developed in other countries. (The Mercedes technique of welded cylinder construction was copied to good effect for the Rolls-Royce Eagle.) Perhaps further development of these types flagged during the war, for the German Air Staff became impressed with the virtues of the light Allied rotaries. (For example, the 'ace' Max Immelman suggested that a Fokker fighter be equipped with an experimental captured Le Rhone for trials.) Oberursel-built copies began to be available from 1917, although it has been said that some of the drawings had to be obtained 'in an underhand way'[5].

The noted aircraft constructor, Fokker, was also convinced of the merits of the rotary for conferring good performance on his aircraft. He acquired the Oberursel company, after subjecting the shareholders to 'a stirring, patriotic address'. They were impressed, it is said, by his sincerity and patriotic motives (though shortly after the war ended, Fokker became re-naturalized as a Dutchman). Oberursel Le Rhones powered aircraft such as the Fokker Triplane and D VI scouts. However pilots preferred the Swedish Thulin Company 'booty' engines[5] for reliability and power. (On test at Farnborough, a captured Oberursel achieved 125 hp at 1300 rpm against a figure of 137 hp from a French-built example.)

The German rotaries also had to contend with inadequate oils due to the shortage of castor (see 'The Castor Question' p.46) and attempts were made to improve mineral oil by exposing it to electric spark discharge. It appears that some castor may have been reserved for 'crack' squadrons.

BESKRIVNING
AV
THULINMOTOR
TYP A

DESS SÖNDERTAGNING OCH
HOPSÄTTNING
SKÖTSEL OCH UNDERHÅLL

The Thulin company made Le Rhone engines under licence in Sweden.

Oberursel U III. A 160 hp twin-row rotary, based on the Gnome 'Mono'. (c 1915)

Rotary Manufacturers after the War

It is remarkable that few of these makers retained any significant aero engine business after the war. In Germany, of course, manufacture of high power engines was prohibited. In Britain, the major contractors for rotaries were general engineers and machinists rather than engine makers, and they returned to their normal business, particularly in view of the abruptly severed orders for military equipment. In France, the appalling casualties that had been suffered brought a widespread revulsion against war profiteering, and it was resolved to impose swingeing taxes on the *marchands de canons*. For this reason, Clerget was no longer able to maintain a position as a major engine maker (though his earnings, it is said, had not gone for personal gain, but to build a new modern plant which had to be sold to meet the tax). He continued in the inter-war period with experiments on aircraft diesel engines. The economic position of Gnome-Rhone can have been no better. After producing their own successful designs in very large numbers, with many more engines being built by their licensees, they were unable to develop successful replacements. Ironically, the great Gnome-Rhone concern became itself a licensee, acquiring from Britain in 1921 the rights to build the Bristol Jupiter.

'Ingénieur génial, passionné de mécanique'. Pierre Clerget with one of the stationary radial diesel engines he developed after the First World War.

War Service and the Demise of the Rotary

It is always interesting to speculate on the contributory factors when, almost overnight, an important class of machine becomes replaced by a succeeding type. Design work on rotaries virtually stopped in 1919 and thereafter, attention passed to stationary engines – principally air-cooled radial and liquid-cooled in-line types. A variety of explanations have been advanced and frequently repeated as to why the once pre-eminent rotary so rapidly became obsolete. The reasons usually given are: a limitation on rotational speed due to centrifugal stress; power loss due to the windage of the rotating cylinders; the gyroscopic effect on the aircraft during turns, which got worse as more powerful (and increasingly heavy) rotaries entered service; and, finally, the high consumption of fuel and oil.

This seems like a comprehensive enough list of drawbacks! However most of them had been inescapable features of the engine even in pre-war days, when T.O.M. Sopwith, the British pioneer aviator had described the 'second generation' Gnome, the Monosoupape, as 'one of the greatest single advances in aviation'. In fact towards the end of the war, Germany, which had concentrated more on the development of efficient stationary engines and had few indigenous rotary designs, started to place emphasis on the production of light rotary engines, and introduced outstanding 'Scouts', the Siemens-Schuckert D III and D IV, powered by a newly developed Siemens

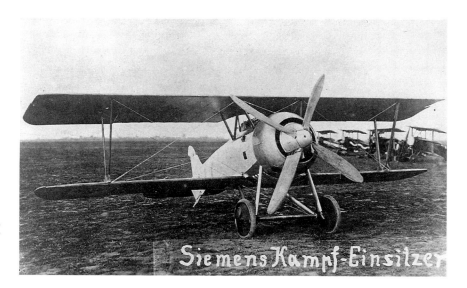

Siemens-Schuckert D III, powered by the contra-rotating Siemens-Halske Sh III engine. 'The best German fighters to reach operational status.'

Sopwith Snipe with 200 hp Bentley BR 2. Although it saw little action during the war, the Snipe was the ultimate Allied rotary-engined scout.

rotary engine. On the Allied side, the most powerful rotary developed, the 200 hp Bentley BR 2, was in front-line service right up to the end of the war, and was invaluable in equipping aircraft like the Sopwith Snipe, which had been originally designed for the fixed radial ABC Dragonfly.[6]

The problem of the 'centrifugal stress' imposed by rotation is frequently cited as a limiting factor in rotary engine development. This does not hold up to examination, for combustion pressure makes up the major contribution to loads on the engine components. A calculation for the stress in the cylinder wall of a 50 hp Gnome is given in a subsequent section.

An article written after the war, entitled 'British Aero Engine Achievement; a Record of Wonderful Progress', noted of the rotary:

> In 1914 it was still in the teething stage and suffered many juvenilities. Its cylinders distorted so grievously that flexible piston rings were employed to 'follow up' the distortion. French genius gradually eliminated its weaknesses. The later Monosoupapes, Clergets and Le Rhone were disfigured only by the shortcomings native to the type – the waste of 10% of the power in rotating the engine body, the high oil consumption, due to the impossibility of circulating the lubricant, and the high petrol consumption . . . Nevertheless the rotary claims a large share in the fighting supremacy of the Allied Scouts. *If limited to 3 hour tankage it was lighter than any available rival*, while its compactness facilitated the design of scouts with powers of lightning manoeuvre.

The effect of gyroscopic forces on aircraft control has often been cited as a major reason for abandoning the type, but although the effect became worse as increasingly powerful rotaries with greater rotating mass entered service, skilled pilots adapted to this idiosyncracy, and even turned it to their advantage.

Sopwith Camel – 'maid of all work' for air fighting in the second half of the war. (The 130 hp Clerget was most usually fitted though some late examples had the 150 hp Bentley BR 1.)

. . . it was just this instability that gave Camels their good qualities of quickness of manoeuvre. A stable machine had a predilection for normal flying positions . . . whereas a Camel had to be held in flying position all the time, and was out of it in a flash. It was nose light, having a rotary engine weighing next to nothing per horsepower and was rigged tail heavy so you had to be holding her down all the time . . . With these unorthodox features, a Camel was a wonderful machine in a scrap . . .'

(Victor Yeates, *Winged Victory*)

It seems reasonable to suppose that as long as the rotary offered clear advantages for combat aircraft, particularly in terms of power to weight ratio, it would continue to be employed in spite of its objectionable features. The 'three-hour tankage' referred to was normal for patrol duty by single-seater scouts, and stationary engine types had similar endurance. The data presented in Table II shows that for scouts of around 200 hp (approximately the upper limit for single-seater aircraft employed in this role in

Australian air mechanics retrieve a Sopwith Camel at Minchinhampton. The undercarriage appears to have collapsed after a heavy landing.

1918) tankage for fuel and oil in both fixed and rotary engined types is remarkably similar, as is the endurance and all up weight of the aircraft. This is not to deny that the rotary was becoming relatively less competitive towards the end of the war, but the point should be made that the two types were more evenly matched, for specific duties, than is often asserted. The

Air mechanics at Rang du Fliers, 1918. Useful parts are being salvaged from damaged aircraft. In the upper photograph, a Le Rhone engine has been removed.

In spite of tricky handling, Camels were used for a great variety of experimental tasks. *top* Take off from HMS Pegasus, 1918.
below 'Alighting Trials' on HMS Furious.

Camel swung into position under the airship R 23. It was flown from the airship by Lt. R.E. Keyes.

production figures for the two principal British scout types in the last years of the war show that in 1917, 1325 Sopwith Camels (rotary engined) were produced, as against 825 S.E.5A's (stationary water-cooled Hispano engine). In 1918, the position had just reversed, with 4377 S.E.5A's made, though Camel production, at 4165 units were not far behind. In all, 5490 Camels and 5205 S.E.5A's were built.

If the war had dragged on into 1919, it seems likely that the rotary would have continued to make a major contribution to the air strength of both sides. However, there would probably have been a reduction in the relative numbers of rotaries for front-line scouts, as a number of aircraft intended to have 300 horsepower engines were at the design stage, and there do not appear to have been any rotaries under development that could have been considered as practical alternatives to the stationary engines that were being perfected in this power class.

The nemesis of the rotary did not arrive in 1918, as is often asserted, and it was not yet obsolete for war service by then. It was obsolescent certainly, in 1918, but it was the end of the war, more than any other single factor which led to the rapid extinction of the type, for thereafter, the pursuit of a high power to weight ratio, almost without regard for other factors, became less crucially important. There were also design constraints which would inevitably have led to the replacement of the rotary in time, even had the war continued, but these have not hitherto been identified sufficiently accurately. (There is, in some respects, an attractive analogy between the role of the rotary in the First World War and the high-power piston engine, vis à vis the turbojet, in the Second World War. Both were almost completely replaced in a very short time after the end of the respective wars).

The Limits to Rotary performance

A study of performance figures drawn from contemporary sources (Tables 3a and b) gives a revealing picture of the comparative merits of rotary and fixed engines during the war. The order in which the engines are listed is not strictly chronological, but it reflects the general design age, and the trend of development during the war. The power to weight ratio in early rotaries is around 3lb/hp for types like the Gnome 80 hp or Clerget 9Z, and though this falls to something over 2 lb/hp quite rapidly, thereafter there is no consistent pattern of improvement through development. By contrast, there is consistent progress in improving the power to weight ratio of stationary engines. It is instructive to look at the Rolls-Royce Eagle series, where development of a single basic type produces a striking improvement in power output. The difference in the development potential seems clear, but the phenomenon requires explanation.

The suggestion advanced here is that because of its special mode of operation, the rotary is effectively limited by a ceiling to the brake mean effective pressure (BMEP) or 'torque' it can develop. Brake mean effective pressure is an indication of the average working pressure being developed in the cylinders, and is, therefore, one measure of the effectiveness of the engine design. Among other things, it reflects how well the cylinders are being filled with fresh mixture before each combustion stroke. As described in the section 'The Rotary Engine Operating Cycle' (p.42), a feature of all the major European rotaries is that fuel and air are drawn into the engine through the hollow stationary crankshaft. The mixture emerges into the crankcase between the crank webs, and passes around the big end assembly and the whirling connecting rods before finding its way, eventually, to the manifolds (or transfer ports in the case of the Gnome 'Mono') which convey it to the combustion chambers. Thus the induction route for the rotary is both long and extremely tortuous – an irremediable disadvantage for the improvement of power output. By comparison, the inlet tracts of stationary aero engines, even of 1918 vintage, are simple and direct. The brake mean effective pressures developed by the engines, shown in Tables 3a and b, reflect this important difference and parallel the power to weight figures already referred to. Quite early in the war, the BMEP figure improves from 62 lb/sq in for the valve in piston 80 hp Gnome, to figures of around 80 lb/sq in. The best result from a rotary (100 lb/sq in) is achieved by the Bentley BR 1. By contrast there is a steady progress in stationary engines, with BMEP's continuing to rise. This implies that volumetric efficiency is improving through development, in spite of the fact that increasing engine revolutions and piston speed tends to make good filling of the cylinders difficult. Again, reference is made to the interesting data from the Rolls-Royce Eagle series, which shows an increase in power between Marks 1 and 8 of over 40 percent (from 254 to 359 bhp), from the same cylinder capacity. This is accompanied by an actual reduction in specific fuel consumption. The improvement in BMEP from 90 to 127 lb/sq in also

The water-cooled Rolls-Royce Eagle – a new standard in reliability and power.

Vickers Vimy with Rolls-Royce Eagles – the aircraft that successfully crossed the Atlantic in 1919.

reflects the increase in efficiency through development. (The V8 Hispano-Suiza shows a similar pattern.)[7]

Although for the reasons outlined, it appeared difficult to gain similar improvements in specific output with rotary engines, powers were raised, mainly by the expedient of increasing the swept volume. The climax of this line of development is the nine cylinder Bentley BR 2 rotary, which developed a maximum of 230 hp, from the then massive capacity of 25.3 litres. It seems likely that in this case, the intrinsic breathing problems referred to were compounded by the added problem of filling cylinders of this large size, for the BR 2 developed a BMEP of 92 lb/sq in, rather less than its predecessor, the similar, but smaller Bentley BR 1.

Rotaries were also handicapped in the development of higher power by being designed to run at lower speeds than stationary engines. First World War stationary engines usually ran at 1800–2000 rpm, as against 1200–1300 for rotaries. This low speed was not selected because of any inherent mechanical limitations, but partly to avoid excessive windage losses[8] and partly also to run airscrews at what contemporary theory held to be an efficient speed, since it was difficult to contrive a mechanism to gear down the propeller on a rotary. In general, increasing the engine speed should allow more power to be developed, but in the case of the rotary, the innate breathing deficiency would tend to become more apparent, and it is unlikely that it would have led to a commensurate increase in power. There is the interesting example from the German side of a geared rotary which was designed to run at 1800 rpm, the Siemens-Halske Sh III. In this ingenious design, the crankcase and cylinders rotated at 900 rpm, while a differential gear assembly at the back of the motor caused the crankshaft to rotate in the opposite direction giving an aggregate of 1800 rpm true engine speed. Thus a higher rotational speed was obtained with rather lower windage losses than those of the standard rotary.

A captured example was tested comprehensively at the Royal Aircraft Establishment. The report enables us to compare this 'high speed' rotary

Wolseley Viper – a licence-built version of the 200 hp Hispano-Suiza.

Carburettor and crankshaft intake of the 50 hp Gnome. The restriction to gas flow resulting from the long inlet path can be envisaged.

"MONO"   OPERATION No. 3

Ignition Plugs being removed.

Gnome

Gnome manuals gave detailed instructions for maintenance and repair. The need for cleanliness and accuracy was stressed.

with 1200 rpm types, and also with contemporary German stationary engines. The increased operating speed does allow the engine to develop greater power relative to its capacity than normal rotaries. However, the results imply (Tables 3a and b) that the Siemens is not able to fully exploit the higher speed for though it has a similar swept volume, and runs at a similar piston speed to in-line 6-cylinder engines like the 200 hp Austro-Daimler and 230 hp Benz, the power output (and the BMEP) is inferior. The torque curve also falls rapidly as the Siemens approaches its operating speed, suggesting the onset of breathing difficulties, and the BMEP is only 80 lb/sq in at 1800 rpm.

As in most other rotaries, the Siemens draws fuel/air mixture through the long hollow crankshaft, via the crankcase, and mixture finally reaches the combustion chamber through radial inlet manifolds running from the crankcase. It seems clear that this long gas pathway is the main restriction on performance, and thus the Siemens does not rival contemporary German stationary engines in performance, even though it runs at a similar engine speed.[9]

A point worth noting from Table 4 is that the valve timings on the Siemens are actually wider than on these more efficient and powerful engines, (the inlet and exhaust valves are open for longer periods). This perhaps implies that the designers had recognized a problem in achieving good breathing and attempted to compensate for it. It also gives further weight to the suggestion that an inherent breathing deficiency – the restriction to gas flow presented by the long crankcase induction system – was the main factor in rotaries limiting the improvement of power outputs.

Conclusions

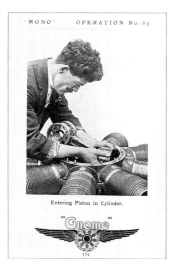

"MONO" .. OPERATION No. 63

Entering Piston in Cylinder.

174

"MONO" .. OPERATION No. 39

Checking alignment of Connecting Rod Bushes.

126

Before the war, the rotary had made an impression on the merits of its performance, although there was a certain amount of confusion about why it worked so well, and an almost magical belief in the particular benefits of rotary motion; (an Adams-Farwell catalogue reads 'The lightest practical powerplant ever constructed, *natural laws* being utilized'). By the end of the war, there was a much more widespread understanding of the type from the viewpoint of internal combustion engineering, and compared to the fast developing stationary engines, its deficiencies were apparent. Not only was the consumption of fuel and oil high, but there was a frequent demand for overhaul (a task that required highly skilled attention), particularly because the engines were built so light, so that cylinder distortion was a constant problem. (Royal Flying Corps fitters' notes refer to engine overhaul life in terms of 'hours efficiency'.) The Gnome Monosoupape was reckoned good for 20 hours against 70–80 for the stationary air-cooled Renault and RAF engines. (By the end of the war, water-cooled Rolls-Royce Eagle engines were able to run for 100 hours between overhauls.) Under wartime conditions these drawbacks can be tolerated, particularly if the alternative power units are heavy and complex and impose a performance penalty on combat aircraft. However the potential of the stationary types could be clearly seen by 1918, and as soon as the end of the war gave a breathing space, virtually all development attention was transferred to them.

It is argued here that the principal defect of the rotary engine, which led to its eventual disappearance, was an inherent restriction of breathing capacity. This was a direct consequence of the method of operation, which required the fuel/air mixture to be aspirated through the stationary hollow crankshaft and the crankcase, rather than through a simple and direct inlet tract, as on contemporary stationary aero engines. This handicap directly affected volumetric efficiency and is reflected in the low BMEP figures recorded for rotaries.

On a volumetric and thermodynamic basis the rotary may have been inefficient, but on broader criteria it was a highly efficient solution to the problems of power output, construction at minimum weight, and reasonable reliability. For a single decade, under the special conditions which obtained from 1908 to 1918, it made a unique contribution to aeronautics.

PART TWO

Technicalities

The rotary engine operating cycle

All the rotary engines mentioned in this book operate on the normal four-stroke (Otto) cycle. To illustrate the functioning of the rotary, it is simplest to look first at the way a basic single cylinder four-stroke stationary engine works.

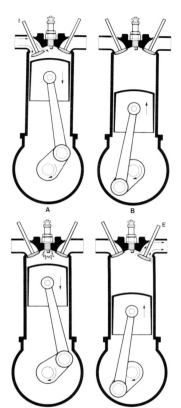

Fig. 1

A The inlet valve I is opened by a cam mechanism, geared to the engine crankshaft. The piston, on its first downstroke, draws in an air/fuel mixture through the inlet pipe.

B The inlet valve closes as the piston reaches the bottom of its stroke. On the following upstroke, the piston will now compress the charge that has been drawn into the cylinder.

C As the piston nears the top of the cylinder, a timed spark initiates combustion of the compressed mixture. The piston is now driven down by the rapid pressure increase of the burning mixture, to provide the working or 'power' stroke.

D As the piston approaches the bottom of its power stroke, the exhaust valve E opens, and burnt gases start to flow out of the cylinder under pressure. This discharge continues as the piston completes its upward stroke. As the piston approaches the top of the cylinder on this stroke, the exhaust valve closes, and the inlet valve opens; the cycle of four strokes is now ready to begin again.

Fig. 1

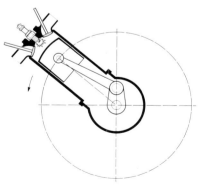

Fig. 2

Now let us consider what would happen if we were to undo the normal mountings of our basic engine and, instead, bolt the crankshaft securely to a test bench (Fig. 2). It should be clear that the crankcase and cylinder would then be free to rotate around the fixed crank, while the piston would still rise and fall in the cylinder in the normal way. The inlet and exhaust valves too would operate normally, and in their correct phasing. If we could provide the spinning cylinder with petrol/air mixture and also ensure that the sparking plug received an ignition voltage pulse at the right moment, the engine would run. In this example of course, the engine would be very unbalanced, and would probably wrench out of its mountings – but it illustrates exactly the way the rotary aero engine operates. In practice, rotaries were built with an odd number of cylinders (usually 5, 7 or 9) radially arranged in one plane around the crankcase, like spokes of a wheel, as in the diagrammatic view of the Le Rhone radial in Fig. 3. This layout gives good mechanical balance, and also provides equally spaced firing intervals. An engine built on this pattern will therefore run smoothly, whether it

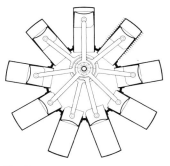

Fig. 3

operates as a 'rotary', spinning round a fixed crank, or as a conventional static 'radial' of the type used extensively in the 1920s and 1930s and during the Second World War. On the rotary, the propeller will be secured to the rotating crankcase itself. On the radial, it will be mounted conventionally, on the crankshaft.

The problem of providing petrol/air mixture to the whirling cylinders of the rotary is solved by using the crankcase as a distribution chamber. The crankshaft is in the form of a hollow tube, and the engine draws air through a carburettor mounted on the inboard end, as in the 50 hp Gnome illustrated in Fig. 4. (In some cases a simple spray jet is fitted in the inlet passage.) The resulting mixture passes along the crankshaft and into the crankcase. From there, mixture reaches the cylinders in different ways, depending on the particular design. In the Le Rhone, Clerget and Bentley

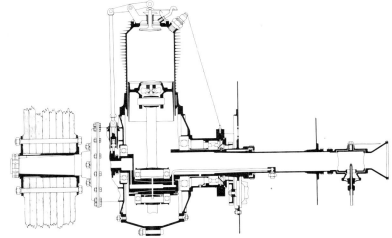

Fig. 4

engines, inlet manifolds (one per cylinder) run radially out from the crankcase to each cylinder head, where admission is controlled by a pushrod operated poppet valve. The cylinder heads of these engines appear fairly conventional, by the standards of the time. However in the original 50 hp Gnome of 1908, the mixture passes directly from the crank chamber through poppet valves in the piston crown. These are spring-loaded and function like the automatic, or 'atmospheric' inlet valves in early motor car engines. When the cylinder pressure falls on the induction stroke, the reduced pressure opens the valve and mixture is drawn in; as compression builds up, they close automatically. These valves are subject to the disadvantage that they open later, and close sooner, than positively actuated ones. They are, in addition, exposed to the combustion heat flowing to the piston crown and as a result, frequently gave trouble.

In the later Gnome design, the Monosoupape, (usually known as the 'Mono') the valve in the piston crown was eliminated, and mixture was transferred from the crankcase to the cylinders by a row of ports, uncovered by the piston at the bottom of its stroke. These are similar to the transfer

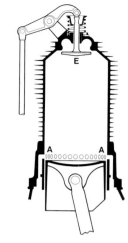

Fig. 5

ports used in two stroke engines, but it should be stressed that the Mono operates on the normal four-stroke cycle, although the timing of the single large exhaust valve in the cylinder head (the 'monosoupape') is unusual.

The sequence of operations is described as follows, with reference to Fig. 5. Imagine that the piston is travelling down the cylinder on its working stroke. The exhaust valve E is opened earlier than is usual, so that by the time the piston uncovers the row of ports AA, the cylinder pressure has fallen and little exhaust gas flows through them to contaminate the crankcase mixture. On the ensuing upstroke, the exhaust valve stays open as usual, and the remaining exhaust gas is discharged. The piston then starts its downward suction stroke, but the exhaust valve is held open for about one-third of the stroke, allowing fresh air to be drawn in. The exhaust valve then closes, and the piston continues its downward stroke, creating a partial vacuum within the cylinder; when the piston uncovers the transfer ports AA, petrol rich mixture from the crankcase is drawn into the cylinder, and mixes with the fresh air already there, to form a combustible mixture. Thus the engine operates on the normal four-stroke cycle, but only one mechanically actuated valve is used. Although the system may seem odd, the BMEP figure for the engine (see Table 3a) shows that it was no less efficient than contemporary rotaries with more conventional valving. 'Monos' were used by the Allied air forces throughout the war.

Control of the Rotary

It is often asserted that the rotary had 'no carburettor' and that power could only be reduced by intermittently cutting out the ignition. This is an oversimplification. In general rotaries did use a simple carburettor which combined a petrol jet and flap valve for throttling the air supply. However this was not an automatic carburettor, as we understand it today; that is, an instrument that will keep the petrol/air ratio reasonably constant over a range of throttle openings. In practice, the pilot would set the throttle to the desired setting (usually fully open), and then adjust the mixture to suit, using a separate lever (the 'fine adjustment') that controlled the fuel valve. This procedure with the Gnome is well described by C. Fayette Taylor; 'Because of the large inertia of the rotating engine, it was possible to adjust to the appropriate mixture by trial, without stalling the engine. After starting the engine with a known setting for idling, the air throttle was opened wide, at which time firing ceased, but rotation continued. The fuel valve was then opened until firing started and maximum propeller speed was attained. Since the reverse process was difficult, 'throttling' was accomplished by temporarily cutting the ignition, keeping the engine going by short 'bursts' of power. Oddly enough, this technique was easy to learn and pilots seemed to like it.'

By the middle of the war, however, some throttling was essential to allow aircraft to fly in formation, and the improved carburettors then in use would allow a power reduction of up to 25 percent. The pilot would close the air valve to the desired degree, and then re-adjust the mixture. Experienced pilots would 'feel' back the fuel lever at frequent intervals, to make sure that the mixture had not got too rich. It was better to err slightly on the lean side, for power recovery would be instantaneous when the fuel supply was increased. By contrast, if the engine was badly over-rich, it could take up to seven seconds to recover, due the large crankcase volume and the possible presence in it of raw petrol. Over-rich running could also cause sparking plugs to soot up, and cylinders to cut out.

An exception to this general control system was the Gnome Mono. Since much of the air for combustion was drawn in through the (cylinder head) exhaust valve, throttling the crankcase air would have been ineffective. Monos, therefore, had only a single petrol regulating control, and this was used for a limited degree of speed regulation. Early examples did also feature variable valve timing, to give greater control, but this was found to cause burning of the valves and was abandoned.

On later rotaries, 'blipping' the ignition was still required for landing. Some engines were equipped with a 'blip-switch' that cut out only a proportion of cylinders, probably to ensure that the engine would keep running, and did not oil up. However the simple ignition cut-out was most common. By 1918, a Clerget handbook counselled, 'Although a rotary engine, the Clerget is not to be treated as though it were a Gnome. Disregard the advices on switch control written at a time when the Gnome engine was considerably used. All the necessary control is to be effected with the throttle. Except in cases of emergency or when a pilot instantly needs his engine at full power – as when making a difficult landing for example – the engine is to be stopped and started by turning the petrol off or on. Use the throttle as much as possible and the 'cut out' switch as little as possible. Continual 'buttoning' with an open throttle is ultimately damaging to the engine.'

The Castor Question

Castor oil was thought 'as indispensable for aero engines as it was for young children'! Contemporary advertisements.

One of the hoariest 'rotary myths' concerns the use of castor oil. Various authors have claimed that castor was essential because, in rotaries, fuel/air mixture passes through the crankcase and that since castor is 'not soluble' in petrol it would not wash off bearing surfaces like a mineral oil. These writers are obviously unaware of the very high outputs attained by two-stroke racing motorcycle engines where the oil is fully soluble in petrol and is introduced into the crankcase with the fuel, whether pump-metered, or as was common until recently, in an actual 'petroil' solution. Conversely, it is also often wrongly maintained that racing cars and four-stroke motor-cycles from the 1920s to the 1950s which used alcohol fuels had to use castor because it *is* soluble in alcohol!

The question of solubility is clearly a red herring. Castor oil was the natural choice for rotaries because it was (and still is) a superb lubricant, and was unquestionably superior to the mineral oils then available. In laboratory tests for lubricity (the property of keeping rubbing components apart under load), castor is still superior to current modified mineral-based engine oils. Its survival in front-line motor sport virtually up to this day has been due to this property. However it has not proved feasible to devise additives to give castor the other qualities now essential for oils such as detergency, long term stability, and the ability to inhibit corrosion. Nevertheless the oil film provided by castor is exceptionally tenacious (partly an effect of carboxylic acid radicals which create a polar affinity between the oil molecules and the metal) and it excels where high loads and temperatures tend to break down hydrodynamic (fluid film) lubrication and only residual boundary lubrication can prevent scuffing and seizure.

These are conditions which would be met almost routinely at the piston/cylinder surface in rotaries, and which specifically made them extremely sensitive to oil quality. By contrast, the crankshaft, big end and cam gear assembly probably posed few lubrication problems even though the oil was simply 'provided' rather than pressure-fed.

Castor, therefore, was the natural choice, especially under the arduous operating conditions of war service. However, mineral oils could be used in rotary engines – particularly in later designs where there had been some development in the control of thermal distortion of cylinder barrels and pistons, and of the combinations of metals used for rubbing surfaces. In the United States, where there was a greater variety of petroleum crudes, and more familiarity with mineral oil, it was routinely used. C. Fayette Taylor recalled 'One of my first assignments in aviation (as Officer in Charge of the US Navy Power Plant Laboratory in 1917) was to make tests to show that mineral oil could be used in aero engines. Previous to that time, castor oil had been considered as indispensable for aero engines as it was for young children'.

The Stress on Components imposed by Rotation

The suggestion has frequently been made that rotaries were reaching mechanical limits because centrifugal loads compounded the normal operating ones. It is, however, unfounded and contemporary designers do not appear to have been concerned about this aspect of the engine. In his authoritative book *Aero Engines* (published in 1916) G.A. Burls showed that the stress on a transverse section of a 60 hp Gnome cylinder barrel is quite moderate, in spite of the extreme thinness of the wall (1.5 mm, or less than one sixteenth of an inch). In this example, the maximum gas pressure in the cylinder contributes a stress of 6000 lb/sq in to the wall, while that due to centrifugal force is only 2100 lb/sq in. The total stress of 8100 lb/sq in is not excessive, especially as these cylinders were machined from nickel chrome steels that would correspond to today's 50–60 ton medium/high tensile steels.

Cylinder head of the Siemens-Halske Sh III. Mechanical stress caused by rotation was not a problem, but all rotaries required balance weights on the rockers to prevent the mass of the pushrod holding the valves open.

Swan-song

Joy-riding and Flying Circuses

Rotary-engined aircraft disappeared rapidly from military aviation following the end of the First World War, though Sopwith Snipes with Bentley engines were retained by the RAF for a few years. Two squadrons were despatched to Turkey in 1922, and another (No 1 Squadron) flew Snipes in Iraq as late as 1926.

However there were large stocks of redundant military aircraft and these were offered for disposal by the authorities at very low prices. The availability of cheap aircraft, and the desire of many former wartime pilots to

Sopwith Snipes being started by Hucks starter (modified Model T Ford).

A Snipe of No. 25 Squadron over Constantinople, 1923.

find some way to continue flying led to the formation of 'joy-riding' outfits. In Britain, a favourite aircraft for this type of work was the Avro 504K, fitted usually with the 110 hp 9-cylinder Le Rhone rotary. Originally designed as a two seat trainer, the rear cockpit could be extended to carry two, or even three passengers.

Public awareness of aviation had been promoted by reports of the new role of aircraft during the war. Books like *Five years in the Royal Flying Corps*, by J.T.B. McCudden, and Major W.A. Bishop's *Winged Warfare* (sub-titled *Hunting the Huns in the Air*!) enjoyed a ready sale and stimulated

Starting a Snipe the hard way!

Snipes of No. 56 Squadron:
Abu Sueir, Egypt, 1920.

The lone Avro in a field. 'Be up
to date and aviate' read the
board displayed by the
Cornwall Aviation Company.
To fly between venues, the
rear cockpit was loaded up
with equipment and Captain
Phillips' two small boys sat on
top. Outside, step ladders and
gear were lashed on.

considerable interest in the exploits of the 'Aces'. The joy-riders were to
capitalize on this interest, and gave many people their first actual sight of
an aircraft. These itinerant flyers toured the country, landing in fields by
pre-arrangement with local farmers. The ground crew would travel on
ahead by lorry to prepare the field and set up a wind-sock. The arrival of the
airmen would be advertised in the local papers, while the pilots would loop
the loop over the nearby town to drum up interest. At first, they were able
to survive simply by taking passengers up for what would probably be their
first flight. One such concern, the Cornwall Aviation Company started in
1924, and by 1931, the founder, Captain Percival Phillips, was able to claim
that he had personally taken up over 55,000 passengers without mishap.
(By 1937, his figure had risen to more than 91,000.) The standard price was
five shillings for a short flight, though daring customers could pay fifteen
shillings for a loop, or £1 for a spin!

Captain Percival Phillips with G-EBIZ – his favourite Avro 504K. In twelve years, he carried over 91,000 passengers, without injury.

By 1931, aviation was becoming more commonplace, and the days of the lone Avro in a field were over. Most major centres and holiday resorts had been visited, and the joy-riding companies found it increasingly difficult to survive by taking up passengers. Thus the 'Flying Circuses' developed, which in addition to the old stand-bys of passenger flights and aerobatic shows, revived their popular appeal by offering a vaudeville mixture of 'crazy flying', wing-walking, 'flying clowns' and the mock bombardment of escaping criminals with flour bags. Even stunts like these, combined with picking up handkerchiefs with the wing-tips, looping through bridges of bunting and parachute drops, could not sustain the commercial viability of these shows, and in 1935, one of the best-known aviation personalities, Sir Alan Cobham, sold the largest of the circuses. The appeal of these displays had inevitably declined as flying became more familiar to the public, and

Landing back with four passengers.

Captain Phillips off Margate. Photographic evidence helped him beat a charge of low flying!

airfields and air transport became commonplace. However the joy-riders and circuses had given many people their first experience of aviation, and had played a valuable part in fostering 'air-mindedness'.

Martin Hearn 'wing-walking' on an Avro 504K belonging to Aviation Tours, probably 1932.

Towards the end of this period, the rotary engine became increasingly rare as newer aircraft were employed, or the old Avros were fitted with later static radial engines like the Armstrong-Siddeley Mongoose or Lynx. At the end of 1935, the Air Ministry refused to renew the Certificate of Airworthiness for rotary-engined aircraft, and the rotary engine became a museum piece. At the time of writing, a handful fly occasionally in Britain under the less stringent provisions of the Civil Aviation Authority's Permit to Fly. The Shuttleworth Trust operates an Avro 504K and a Sopwith Pup, both with Le Rhone engines, and a 1912 Blackburn monoplane, powered by a 50 hp Gnome. A few other aircraft are in the hands of individual collectors.

Notes

1. Among the interesting solutions adopted by the gifted designer of the Antoinette, Leon Levavasseur, was a form of continuous fuel injection to each valve port. He avoided the usual heavy iron water jackets cast integrally with the cylinders by adopting, in some versions, light copper jackets. These thin shells of pure copper were electrolytically deposited onto the cylinders over a former built up from wax, blackleaded to give the surface electrical conductivity. The wax was subsequently melted out.

Though the Antoinette was designed originally as an aviation motor, examples were adapted for motor boat racing, presumably to recoup some of the costs of the aeronautical experiments.

2. An unusual feature of this engine is the presence of auxiliary exhaust ports in the cylinders, uncovered by the piston at the bottom of its stroke. These were sometimes used in contemporary static engines, like the Anzani, and helped reduce the heat load on the poppet exhaust valve – a critical component in this period, due to inadequate high temperature alloys. These auxiliary ports were abandoned in production Gnomes.

3. Particularly 'Y alloy' developed at the National Physical Laboratory. Shortly after the war Air Commodore 'Rod' Banks worked at Peter Hooker Ltd of Walthamstow which had built many Gnome and Le Rhone engines under licence. He recalled

> 'Our General Manager, Hedley Thomson had a good brain but, like a number of brilliant people, often went off at a tangent. He thought Y alloy, which was largely his baby was the universal material, and could be used for most parts of an engine . . . he decided to try a Gnome cylinder in this material . . . When I said it would just fly off he became impatient, got out his 20-in slide rule and told me that the working stress was well within the limits of the material. I tried to tell him that . . . [the design] . . . limited the dimensions of the Y alloy cylinder to those of the steel one. He wouldn't listen however. . . . I advised the tester not to put the engine in the usual escargot cowling, since the cylinder would soon come off. It was sufficient, I said, to try it in the open . . . but to keep clear. Sure enough, when the Y alloy cylinder fired, before the engine had completed a revolution, off came the cylinder, sailing about 70 feet over the test house . . . we heard no more of Y alloy cylinders.'

4. The scout was what today would be called an 'air superiority' weapon. A major task was to deny the enemy the use of reconnaissance and artillery-spotting aircraft over Allied lines, and to protect Allied aircraft engaged in similar work. Thus the fighting scouts of both sides had the highest performance in terms of speed, manoeuvrability, and 'time to height' of all aircraft in the conflict. They also engaged in some 'ground straffing' for infantry support. Towards the end of the conflict, multi-engined long range bombers made their appearance. These invariably had stationary engines, and it is not suggested that rotaries could have been employed advantageously in this role.

5. The Swedish Thulin company was an official Le Rhone licensee, and is alleged to have supplied good numbers of its well built engines to Germany. These appear to have been marked with name plates declaring them to be captured *(Beute)* to conceal their origin. It is possible that the Thulin company was also the source of drawings and specifications for the Oberursel copy. Thulin always denied supplying Le Rhone engines to Germany, but the profits of the company declined from 1.25 million kroner in 1917, to just 16,000 kroner in 1918.

6. This affair was something of a debacle. The Dragonfly was selected for production by the British authorities on the basis of an untried 'paper' design, and on the success of an earlier engine from the designer Granville Bradshaw. A substantial commitment was made for its production (over 11,000 were ordered, from 13 separate manufacturers) but on test at Farnborough, the first examples were found to overheat, to suffer from crankshaft torsional vibration problems, and failed to reach their designed power output. It was fortunate that it was possible to substitute the proven Bentley BR 2 rotary, for which production facilities existed, or a serious shortfall in combat aircraft reaching France would have occurred, particularly if the war had continued into 1919.

7. The Hispano-Suiza V8 engine was licence-built in Britain as the Wolseley Viper. A great deal of development effort was devoted to extracting what were then very high outputs. With these engines, the Allies began to catch up and overtake the German lead in efficient stationary engines. A 1917 German report commented, 'Among the foreign aircraft engines recently captured by us, after the Rolls-Royce, the Hispano-Suiza undoubtedly merits our interest in the widest sense'.

8. Windage losses amounted to around ten percent and S.D. Heron has recorded 'rotary engine builders objected strenuously to the power of their engines being determined on the dynamometer. They insisted that torque stands should be used. On a torque stand, the very considerable windage loss due to rotating the cylinders was recorded as useful horsepower.' It should be noted that stationary water-cooled engines also incurred a 'power debt' through the drag induced by their radiators. This, of course, would affect aircraft performance, though not dynamometer readings.

9. The scouts equipped with the Siemens-Halske geared rotary did have excellent performance (particularly in rate of climb). Whether the extra mechanical complication was worthwhile could be questioned, and early service examples of the Sh III were plagued by piston seizures which resulted, perhaps, from the reduced cylinder rotational speed. Though the Sh III was the most powerful German rotary, the larger Bentley BR 2 was simpler and more powerful, and it could be argued that W.O. Bentley's design solution was more economical and effective. The leading particulars of the Siemens can be compared with those of the BR 2 in Table 3a.

Appendix

Winged Victory

Flying a rotary engined scout in the First World War.

Victor Yeates' much acclaimed autobiographical novel, *Winged Victory*, first published in 1934 by Jonathan Cape, has been noted for the accuracy with which it described front-line life with a squadron during the First World War. This extract on learning to fly a Camel is included because it illuminates vividly the problems and advantages of the rotary engined scouts.

'I suppose you haven't run a Clerget engine before' (It was a Clerget Camel) . . . 'You'll find it just like a Le Rhone. . . . 'You'll find it a bit fierce to start with: you've got another forty horse-power and plenty more rev's. You'll soon get to like that. . . . Be careful with your fine adjustment, they're a bit tricky on that . . . I expect you've heard all about flying them. Be careful of your rudder. You may find it difficult to keep straight at first. Keep a shade of left rudder on to counteract the twist to the right; when you're on anything like full throttle you can feel the engine pulling to the right all the time. Remember to use the rudder as little as possible, you hardly want any when you turn. But don't be afraid of putting on plenty of bank. A Camel's an aeroplane, not a house with wings, and you can put 'em over vertical and back again quicker than you can say it.'

Camels of 32 Squadron at Humières, April 1918. A Bristol Fighter is overhead.

Camels over the Royal Naval Air Service aerodrome at Chingford.

RNAS Camels flying over two Curtiss Jenny trainers and an Avro 504K.

Tom had got in and run the engine. There wasn't any difficulty about that. He taxied out and turned round. He opened the throttle and the engine roared. Then it spluttered. Hell! Too much petrol. His hand went to the fine adjustment. By the time he had got the engine running properly he was almost into a hangar . . . he pulled the stick back and staggered into the air just clearing the roof: if the engine gave one more splutter he would stall and crash. But the engine continued to roar uniformly. His heart, having missed several beats, thumped away to make up for them, and he felt emulsified; but he was flying.

The engine was pulling like a chained typhoon. He seemed to be going straight up. Two thousand feet, and he had only just staggered above the hangars! It was difficult to hold the thing down at all . . . He soon became aware that he was not flying straight. A first, the sensation peculiar to sideslipping had been lost in the major sensation of flying a strange machine, but when his senses were less bewildered by the strangeness of it he became aware of a side-wind . . . of the particular feeling of wrongness that is associated with side-slipping. He had seen beginners doing that sort of thing. A few days previously someone had taken off on a Camel and gone across the aerodrome almost like a flying crab, while everybody held their breath and waited for the side-slip to become a spin and the pilot a corpse; but he had got away with it. Tom had been scornful, but here he was, doing the same sort of thing; he had no idea why. He could fly any ordinary aeroplane straight enough. He experimented with the rudder, but soon came to the conclusion that sideslipping was an ineluctable vice of Camels; at any rate of this one. It would not fly straight for more than a second at a time. At five thousand feet, he put the machine on a level keel in order to try to turn. . . . Then, very carefully, he pressed the stick towards the left and the rudder gently the same way. What happened was that all tension went out of the controls, there was an instant of steep side-slip, and the earth whizzed round in front of him. A spin! At once, his hand went to the fine adjustment

to shut off the petrol. . . . For some minutes he didn't care to do anything except fly as straight as he could, and it cost him an effort of will to decide to try again. This time he was ready for a spin, and as soon as he felt the controls going soft he came out of the turn. By this means he succeeded in turning through a few degrees without actually spinning, and after a few more such turns he let his strong desire to get back to earth have its way. He made out that he was some way east of Croydon, and it was necessary to turn west. To do this he shut the engine off and brought the machine round in a long sweeping glide. The thing would turn on the glide without spinning, anyhow: that was something. . . . He wouldn't have stopped up any longer for all the wealth of Jewry.

The prototype Sopwith Camel in the snow at Brooklands, December 1916.

But that was long ago: four months in fact. Or was it four years? Camels were wonderful fliers when you got used to them, which took about three months of hard flying. At the end of that time you were either dead, a nervous wreck, or the hell of a pilot and a terror to Huns, who were more unwilling to attack Camels than any other sort of machine . . . A Camel in danger would do the most queer things . . . and in the more legitimate matter of vertical turns, nothing in the skies could follow in so tight a circle . . .

The same with the half-roll. Nothing would half roll like a Camel. A twitch of the stick and a flick of the rudder and you were on your back. The nose dropped at once and you pulled out having made a complete reversal of direction in the least possible time . . .

When Tom had arrived in France for flying duty, not feeling at all sure of his ability to fly a Camel even moderately well, his ears feverish with rumours of enormous casualties among Camel pilots, he was sure his life wasn't worth two sous. Bold, bad, terrific Huns would pounce on him like hawks on a sparrow. But when he got to his squadron he was surprised to find everyone as happy as cats in a dairy. There was no wind up about Camels at all.'

Table 1 Production figures – Société des Moteurs Gnôme

1908	3
1909	35
1910	400
1911	800	(Net profits £148,934)
1912	1000	(Net profits £210,326)
1913	1400

Table 2 Performance figures for some late WWI rotary and stationary engined scouts

	Speed at 10,000 feet (mph)	Climbing Performance: Time (in minutes) to various heights			Endurance (hours)	Tankage (gals)		Aircraft weight loaded (lb)	Engine
		5,000 feet	10,000 feet	15,000 feet		Petrol	Oil		
Rotary Engine									
Sopwith Camel	113	5	10.6	20.7	$2\frac{1}{2}$	26	$5\frac{3}{4}$	1524	Clerget 130hp
Sopwith Snipe	121	3.75	9.4	18.8	3	$38\frac{1}{2}$	7	2020	Bentley BR 2 200 hp
Siemens DIII	112.5	(3.75 mins to 6560 ft)	(6.2 mins to 9840 ft)	(13 mins to 16,400 ft)	2	–	–	1595	Siemens Sh. III 200 hp
Fokker D6	122.5		(9 mins to 9840 ft)	(19 mins to 16,400 ft)	–	–	–	1283	Oberursel (Le Rhone copy) 100 hp
Stationary Engine									
Fokker D7	116	(8.3 mins to 6560 ft)	(13.5 mins to 9840 ft)	(31 mins to 16,400 ft)	$1\frac{1}{2}$	–	–	1936	Mercedes in-line 6 cyl. 160 hp
Sopwith Dolphin	121.5	5	12	23	$2\frac{1}{4}$	27	4	1959	Hispano 200 hp
S.E.5A	125	(6 mins to 6560 ft)	10.3	18.8	3	35	3.5	1953	Hispano 200 hp

Table 3a   Rotary Engine Performance

Engine	BHP at normal revs	BMEP at normal revs (lb/sq in)	Normal revs	Piston speed (ft/min)	Fuel consumption (lb per normal hp per hour)	Dry weight (lb)	Dry weight per normal hp (lb)	Swept volume cu in
Rotaries								
Gnome 80 hp	65	62	1150	1056	0.736	212	3.26	721.2
Clerget 7Z	82	74.6	1300	1180	0.585	234	2.85	722.75
Clerget 9Z	122	77.7	1300	1312	0.79	367	3	992.25
Clerget 9B	134	85.4	1300	1312	0.585	385	2.87	992.25
Le Rhone 110 hp	137.5	86.7	1250	1449	0.596	336	2.56	919.4
Gnome Mono 150 hp	154	96.7	1300	1449	0.82	320	2.08	970
Bentley BR 1	158	100	1250	1394	0.625	408	2.58	1053
Clerget 11EB 200 hp	197	82	1300	1620	0.79	512	2.6	1440
Siemens & Halske Sh.III	205	80	1800	1653	0.65	460	2.24	1134.1
Bentley BR 2	230	92	1300	1536	0.63	498	2.165	1521

Table 3b Stationary Engine Performance

Engine	BHP at normal revs	BMEP at normal revs (lb/sq in)	Normal revs	Piston speed (ft/min)	Fuel consumption (lb per normal hp per hour)	Dry weight (lb)	Dry weight per normal hp (lb)	Swept volume cu in
Renault 80 hp (V8)	102	81.7	1800	1536	0.66	480	4.7	548
RAF 4A (V12)	163	89	1800	1653	0.603	680	4.17	806
Salmson radial (9 cyl)	140	98.7	1250	1148	0.54	555	3.97	897
Curtiss OXS V8	92	111.4	1300	1083	0.46	380	4.13	502
Hispano-Suiza 150 hp (French, V8)	165	114	1600	1365	0.5	440	2.66	716
200 hp Austro-Daimler (in line 6)	200	123	1400	1607	0.55	728	3.64	916
230 hp Benz (in line 6)	230	113	1400	1744	0.63	863	4.33	1148
Wolseley Viper (Hispano)	220	121	2000	1708	0.52	600	2.27	716
Rolls-Royce Eagle 1 (V12)	254	90	1800	1950	0.56	900	3.54	1239
Eagle 5	307	108	1800	1950	0.53	910	2.96	1239
Eagle 8	359	127	1800	1950	0.5	926	2.58	1239

Table 4 Inlet Valve Timing. Siemens & Halske compared with contemporary German stationary engines

	Opens	Closes	Total	BMEP (lb/sq in) at 1600 rpm	BMEP (Max)	Max BHP	Capacity (litres)
Siemens & Halske Sh.III (rotary)	20° BTDC	70° ABCD	270°	90	100 (at 1000 rpm)	205	18.6
200 hp Austro-Daimler (in line 6 cylinder)	5°	20°	215°	120	123 (at 1400 rpm)	222	15
230 hp Benz (in line 6 cylinder)	10°	55°	245°	107	119 (at 1100 rpm)	250	18.8

Table 5a   Production of rotary and stationary aero engines for British Forces 1914-18

	Rotary	Stationary water-cooled	Stationary air-cooled	Total (all types, by year)
1914	46	55	29	130
1915	917	656	1041	2614
1916	2412	1822	2969	7203
1917	5830	7289	3486	16,605
1918	10,090	18,160	2995	31,245
Total (by type)	19,295	27,982	10,520	57,797

Table 5b   British output of rotary and stationary engined single-seat scout aircraft 1914–1918

	1914	1915	1916	1917	1918	Total
Rotary Engined						
Bristol Scout	12	123	236	13	–	384
Martinsyde Scout	11	58	–	–	–	69
Sopwith Pup	4	32	64	973	733	1806
Sopwith Camel	–	–	–	1325	4165	5490
Sopwith Snipe	–	–	–	–	497	497
Total by year	27	213	300	2311	5395	8246 (Total rotary for war)
Stationary Engined						
S.E.5A	–	–	–	828	4377	5205
Sopwith Dolphin	–	–	–	121	1411	1532
Total by year				949	5788	6737 (Total stationary for war)

Production data drawn from *The War in the Air, Vol III* (History of the Great War, based on Official Documents) H A Jones, Oxford 1931

Bibliography

Air Board Tabulated Engine Data, (loose-leaf technical notes on Allied engine types issued in the course of the war). London: Technical Department, Air Board, 1917-8

M.H. André, *Moteurs d'Aviation et de Dirigeables,* Paris: L. Geisler, 1910

G.D. Angle, *Airplane Engine Encyclopaedia,* Dayton, Ohio: The Otterbein Press, 1921

F.R. Banks, *I Kept no Diary,* Shrewsbury: Airlife, 1978

W.O. Bentley ('W.O.'), *The Autobiography of W.O. Bentley,* London: Hutchinson, 1958

A. Bodemer, Histoire des moteurs à pistons de Gnôme et Rhône et de la SNECMA. Pionniers, *Revue Aeronautique des Vieilles Tiges,* 65, (July 1980) pp.5-20

G.A. Burls, *Aero Engines,* London: Charles Griffin & Co Ltd, 1916

C.F. Caunter, *Rotary Engines.* Unpublished manuscript deposited in the Science Museum Library, 1975

T. Chapman, *The Cornwall Aviation Company,* Falmouth: Glasney Press, 1979

A.E.L. Chorlton, Aero Engines, *Journal of the Royal Society of Arts,* Vol.69, (1921) pp.689-705, 707-24, 725-40

P. Gray and O. Thetford, *German Aircraft of the First World War,* London: Putnam, 1962

B. Gunston, *World Encyclopaedia of Aero Engines,* Cambridge; Patrick Stephens, 1986

J.G.G. Hempson, The Aero Engine up to 1914, *Transactions of the Newcomen Society,* Vol.53 (1981-2)

S.D. Heron, *History of the Aircraft Piston Engine,* Detroit: Ethyl Corporation, 1961

H.A. Jones, *The War in the Air, Vol.III,* Oxford: Clarendon Press, 1931

T.M. Mackay, *General Notes on Royal Flying Corps Engines.* Manuscript notes made in the course of the First World War at Blenheim Barracks, Farnborough. Held in the Aeronautical Collection, Science Museum, London

Rear Admiral L.S. McCready, The 50 hp Gnome Omega, *W W I Aero,* (September 1983)

Sir Walter Raleigh, *The War in the Air, Vol.I,* London: HMSO, 1922

Rapport Technique, Commission Interalliée de Contrôle Aeronautique en Allemagne, Chalais-Meudon, 1921

Captain W.H. Sayers, The Development of the Aero Engine Past and Future, *The Aeroplane,* supplements to Vol.16, Nos.23-5 (1919) pp.2321-2, 2415-6, 2509-10 and Vol.17 (1919) pp.33-4

R. Schlaifer and S.D. Heron, *The Development of Aircraft Engines and Fuels,* Boston: Harvard University, 1950

L.J.K. Setright, *The Power to Fly,* London: George Allen and Unwin, 1971

C. Fayette Taylor, *Aircraft Propulsion,* Washington: Smithsonian Institution, 1963

A.R. Weyl, *Fokker: The Creative Years,* London: Putnam, 1965

V.M. Yeates, *Winged Victory,* London: Jonathan Cape Ltd, 1934 (and various reprints)

'XYZ', British Aero-Engine Achievement; A Record of Wonderful Progress, *Autocar* (8 Feb, 1919) p.181

Instruction Manuals and Manufacturers' Handbooks:

The Gnome Engine; its care, maintenance and use, J.D.B. Fulton, London: HMSO, 1913

Notice concernant le Réglage et l'Entretien du Moteur Rotatif Type 9B, 130 hp, Paris, Levallois-Perret: Clerget-Blin et Cie

Manual of Clerget Aero-Motors, London: Air Board, 1918

Care of the 100 hp Monosoupape Engine, London: The Gnome and Le Rhone Engine Co

The Care of the 110 hp Le Rhone Engine, London: The Gnome and Le Rhone Engine Co

Bentley BR 2 Aero Engine. Descriptive Handbook, London: Air Ministry, 1925

Official British reports on captured engines:

Report on the 200 hp Siemens-Halske Aero-Engine (Geared rotary type), London Air Ministry, 1920

Report on the 110 hp German Le Rhone Engine (Oberursel), London: Ministry of Munitions, 1918

Report on the 200 hp Austro-Daimler Aero Engine, London: Ministry of Munitions, 1918

Report on the 230 hp Benz Aero Engine, London: Air Board, 1917

Printed in the United Kingdom for Her Majesty's Stationery Office by Acolortone.
Dd. 238931, C45, 3/87.

Captain Percival Phillips with
his partner A.J. Adams,
engineer to the Cornwall
Aviation Company. The
engine is a 110 hp Le Rhone.